APOGALACTICON
A five act play about the extinction of the dinosaurs

Copyright © 2024 Steven Wickson

All rights reserved

ISBN-13: 9798324960162

Prologue

66 million years ago in the late Cretaceous period the Earth was a very different place. The world was warm and the seas were higher. The polar regions were void of ice caps and instead supported subtropical rainforests. A diverse flora had spread across the landscape that included early flowering plants, conifers and ferns, provided homes for insects, birds, mammals and small reptiles. At the high end of the food chain were dinosaurs, crocodiles and pterosaurs. Powerful reptilian beasts terrorized the landscape while the ancestors of modern mammals trembled in fear. The oceans were very productive and reef environments flourished. The open ocean was populated by fish, sharks, ammonites, sea turtles, Mosasaurs and Elasmosaurs. But something is about to change, for this world was not meant to last. A new era is dawning and what happens in the next few hours will determine what Earth's ecosystems will look like in the ages that follow.

Characters
Tyrannosaur
Dromaeosaur
Ornithomimid
Titanosaur
Ankylosaur
Ceratopsian
Pachycephalosaur
Hadrosaur
Hypsolophodon
Pterosaur
Crocodile
Mosasaur
Elasmosaur
Sea turtle
Shark
mammal
bird
snake
lizard
frog
fish
ammonite
lobster
spider
Red
Green
Blue

Sets
the plain
the forest
the desert
the cave
the open ocean
the reef
the swamp
the refuge

Act 1, Scene 1
Setting: A hot day on the plain. Dinosaurs roam, pterosaurs fly and mammals hide.
Enter: Tyrannosaur, Dromaeosaur, Ornithomimid, Ankylosaur, Ceratopsian, Pterosaur, mammal

Tyrannosaur: I am the best. No one messes with me. Every living thing surrenders to my will. My head is so big and my teeth are so sharp. The puny skulls of all other creatures on this plain can be crushed by the ultimate power of my jaws and feet.
Ornithomimid: Oh no no no no no, the Tyrannosaur is coming. I will have to tell the others.
Ankylosaur: Stay out of my foraging area, Mimus. You always seem to mess things up. You look like you are in a hurry today. What seems to be the problem this time?
Ornithomimid: I came to tell you that Rex, the big, bad Tyrannosaur is coming this way.
Ankylosaur: That may be so Mimus, but unlike you, I do not run from predators, I stand and fight. And Rex will know to back off when he sees my bonecrusher.
Ceratopsian: My ancestors may have hidden from carnivores like Rex. But Ceratopsians like me now have these magnificent headcrests and horns. So I will not bed a knee to Rex.
Ankylosaur: On the other hand, Ornithomimus, may have to run.
Dromaeosaur: I swear I saw something moving around here. Something with hair. Where have you gone little furry creature? Are you hiding under this rock?
Ornithomimid: Dromaeosaur, you should get out of here. The Tyrannosaur is coming.
Dromaeosaur: Don't disturb me right now. I am in hot pursuit of my lunch.
Ornithomimid: Someone else might make you part of their lunch if you don't get out of here soon.
Dromaeosaur: Oh, I get it. You sneaky little thief. You want me to go so that you can have my lunch. Well that's not going to work. Besides, I have a deal with the Tyrannosaur. I eat the little morsels and he eats

the big ones. It's a mutually beneficial arrangement. You could call it symbiotic.

Ornithomimid: Bye. (exit)

Tyrannosaur: Dromaeosaur, you little twerp. What are you doing on my plain?

Dromaeosaur: Oh, I've just been admiring how you are able to dominate the plain and how well you clean up all those big, fat, annoying herbivores.

Tyrannosaur: Well, now I'm just thinking of eating you.

Dromaeosaur: Ah Rex, but don't you remember our mutually beneficial arrangement? We can work together.

Tyrannosaur: Me Carnosaur. Me hungry.

Dromaeosaur: Hey! I thought we were on the same team here?

Tyrannosaur: Hey! I go where I want, when I want and eat everything that gets in my way.

Dromaeosaur: Well in that case I will run away now. Perhaps you can give me a head start?

Tyrannosaur: RAAAR!

Dromaeosaur: Ahhh! (exit)

Tyrannosaur: Everything here is my domain.

Ceratopsian: Keep your distance Rex. That will be as far as you go today tyrant lizard.

Tyrannosaur: Spare me your words, Ceratops. Your fancy face is useless against me. I know that behind that frill you so proudly display, you are weak and fleshy like everything else.

Ceratopsian: Don't test me, big mouth.

Tyrannosaur: RAAAR!

Ankylosaur: Stop tyrant. You might be able to top one of us in single combat but together we stand strong. We are dinosaurs too.

Tyrannosaur: Well now you're just aggravating me.

Ankylosaur: We have had enough of you, Rex.

Tyrannosaur: Ow my toe!

Ceratopsian: Get out of here while only your toe hurts.

Pterosaur: Where have you hidden yourself furry creature?

Tyrannosaur: Ahhh! My toe, my toe.

Pterosaur: Tyrannosaur, you sound like you're in pain.

Tyrannosaur: Ankylosaur clubbed me on the toe and it hurts.

Pterosaur: Ha ha ha. So the big, bad carnivore got smacked around by the plant eaters.

Tyrannosaur: Don't make fun of me or I'll bite your wings off.

Pterosaur: See you later, Rex. (exit)

Tyrannosaur: RAAAR! This land belongs to me. When my toe heals I'm going to make everyone pay for what happened here today. (exit)

Mammal: Whew!

Act 1, Scene 2
Setting: Sunset in the forest. Dinosaurs eat plants, a predator enters forest, and small animals come out at night.
Enter: Dromaeosaur, Titanosaur, Pachycephalosaur, Hadrosaur, Hypsilophodon, mammal, bird, snake

Titanosaur: These plants are so difficult to chew. How do you manage to process them so easily Hadrosaur?

Hadrosaur: I just gather them in my bill here and grind them into a pulpy mush. Yum.

Titanosaur: They aren't like the conifers my kind used to eat.

Bird: Well, I like these flowers because they attract insects, my favourite meal. In return, the insects help the flowers pollinate, attracting them with appealing colors and scents. They make the forest beautiful.

Pachycephalosaur: These flowers taste okay to me. I particularly enjoy the hardy upland varieties. They seem to have a bit of a crunch to them.

Hadrosaur: I like my vegetation to be mixed with a little bit of swamp water. It just makes it a little bit mushier.

Hypsilophodon: I prefer the shrubby plants that don't grow so tall. The rest of you can go ahead and eat the leaves in the forest canopy so the shorter shrubs can get more sun.

Titanosaur: Thank you. I will continue to do that.

Bird: I've always wondered how Titanosaur manages to lift its neck up so high? It must be so heavy. Its brain must not be much bigger than mine, but it must have an incredibly strong heart.

Snake: Sss. While the others ssstomp around, grazing on the leavesss of the forest, I ssslither around in the undergrowth hunting insectsss.

Dromaeosaur: Oh gosh, I am getting hungry. Eggses, where are you hiding your eggses?

Bird: Tweet! Dromaeosaur! What is that egg thief doing here? Although my eggs are small and hidden way up in the trees, he is very cunning. I don't know if the other dinosaur eggs will be safe.

Dromaeosaur: There are definitely dinosaur eggs around here, I can smell them.

Titanosaur: Halt Dromaeosaur. Retrace your steps back the way you came. You are not welcome here.

Dromaeosaur: Relax Titanosaur. I am just passing through the forest. Seeking shelter on this cool evening.

Titanosaur: You don't eat plants. This is not your place.

Dromaeosaur: Uh, well, I have decided to change my ways and switch to eating plants. They are obviously much healthier and they are so tasty.

Hypsilophodon: That's great to hear. You can be part of our group. You are right. There are many advantages to being a herbivore.

Titanosaur: That's nonsense Dromaeosaur. You are a liar and a thief. You don't have a stomach that can digest plants.

Dromaeosaur: Sure I do, it's not all that different than eating meat. My teeth will just slice right through these leaves and stems and give me the nutrition I need.

Hadrosaur: Okay, prove it then. If you think you can eat this vegetation, then let's see you swallow it.

Pachycephalosaur: I'll bet you can't even hold down a mouthful.

Dromaeosaur: Fine. This looks like some nice vegetation over here. (gulp)

Hypsilophodon: Wow, I guess carnivores really can become herbivores then.

Titanosaur: I am not convinced by this charade. Just stay away from my foraging grounds Dromaeosaur or I will stomp on you.

Pachycephalosaur: Yeah! Just because you can act like a planteater doesn't mean that you are one.

Hadrosaur: Come on fellow dinosaurs, if we present a hypothesis, should we not be prepared to accept the results whether we like them or not.

Hypsilophodon: I think it's great. There is hope for carnivores. They can change their ways. Dromaeosaur you are now one of us. Here, follow me and I will introduce you to my youngling.

Dromaeosaur: I will be pleased to meet them.

Titanosaur: No Hypsilophodon. Wait, this isn't right. Dromaeosaur was born a carnivore and will always be a carnivore. He is just mimmicking our ways to get to our eggs.

Dromaeosaur: I am sorry for what I've done in the past. Please do not remind me. It makes me sad.
Pachycephalosaur: I've heard enough of this. (exit)
Hadrosaur: Stop it Dromaeosaur. Dinosaurs don't cry.
Hypsilophodon: It's alright Dromaeosaur. I believe there is good in all dinosaurs.
Titanosaur: No Hypsilophodon. The Dromaeosaur is just pretending. He is not a herbivore and never will be. Any moment now, he is going to barf up those plants and try to eat you.
Hypsilophodon: Oh Titanosaur, don't you see? We can all work together to overcome the carnivores in nonviolent ways. We must embrace and encourage Dromaeosaur as he tries to be a better dinosaur. (Dromaeosaur sneaks away)
Titanosaur: The Dromaeosaur is a trickster and a thief. I've seen the way it acts. It is weak but it is cunning and will do anything it needs to get what it wants.
Hadrosaur: Well, I'm not totally convinced Dromaeosaur is a converted herbivore, but it is quite small. It won't be eating any of us.
Dromaeosaur: Yuck, plants are so disgusting. Mmm, a Titanosaur egg.
Hadrosaur: Hmm. I saw five dinosaurs a minute ago, now I only see three.
Titanosaur: Uh oh.
Hypsilophodon: What's wrong?
Titanosaur: There is a carnivore in the forest.
Dromaeosaur: I sense it is time for me to go now. (exit)
Titanosaur: While we have been debating, the little thief has snuck away. Ahhh, my egg!
Hadrosaur: Oh no.
Hypsilophodon: What happened to your egg Titanosaur? It looks like something broke it open and had an omlette.
Hadrosaur: I am so sorry Titanosaur. I pray that this will not be the end of your great lineage. (sun sets)
Mammal: Ah, the quiet of night. There is peace in darkness. The dinosaurs sleep and I am free to move around the forest.
Snake: Sss, oh now what? Listen you furry runt. This is my territory so stop eating my insectsss.

Mammal: Come on now. The insects have been very active lately and I am sure there's enough to go around. I'm sorry I do not recognize you. Are you new to the forest?

Snake: I am sssnake, you may know my older sibling, the lizard, who is also my progenitor. However, unlike them, I do not have legs and I feel I can move much better without them.

Mammal: Well you can eat the less tasty insects that I don't eat.

Snake: Sss!

Bird: Break it up you two. There are plenty of insects around here. Mother nature encourages them to multiply in number and grow bigger and better to feed us all.

Mammal: That sounds just plain ludicrous to me. Do you really think that mother nature thinks we are the most important component of the ecosystem? I don't think this mother nature being even exists. And by the way, the last time I checked you were a dinosaur, and I hate dinosaurs.

Bird: Oh mother nature exists, I tell you. She's the one that helps the plants grow and reproduce. She tells them to make their flowers pretty and to smell nice so that animals will want to interact with them. Without her the forest would look ugly and bland.

Mammal: That just sounds like dinosaur propaganda. The forest merely looks pretty to us because we grew up in it and it is familiar. If we were from another place, these plants would look alien and weird.

Snake: These plants are ugly and so are fur and feathers. Just leave me alone to slither on the ground and eat my insects. (exit)

Bird: Suit yourself. (exit)

Act 1, Scene 3
Setting: Midday in the open ocean. Sea creatures lurk and predators chase their prey.
Enter: Mosasaur, Elasmosaur, Sea turtle, Shark, fish, ammonite

Fish: I am so happy to be in the open water among the waves. Here, the ocean is my playground.

Ammonite: Yes indeed. It's great out here. I can be the master of my own little world. And whenever things get rough, I can just retreat to my shell.

Fish: There is some nice plankton out here in the open ocean.

Ammonite: Yes, there are a lot of tasty morsels that are easy to catch.

Fish: I don't think you're able to get all the types of plankton that I can catch, with my swift swimming and all. Your shell looks cumbersome and your awkward swimming methods hold you back.

Ammonite: You don't know what you're talking about. I kind swim just as fast as you. As I recall, my kind used to eat things like you back in the Ordovician period.

Fish: Perhaps once upon a time but not anymore. The times have a-changed and we have adapted. Your kind, on the other hand, are still dependent on old techniques.

Ammonite: Don't make me ink on you.

Sea turtle: Hey there, have you two seen any jellies around here? I'm starting to get a little famished.

Fish: Not many jellies around here. Just eat the ammonite. Your strong beak could probably crush its shell, and then I'd have more plantkon for myself.

Ammonite: I don't think so. My shell is designed in a very stable geometric form. Eat the fish, it has lots of soft tissue.

Fish: Your shell is a burden, not a blessing. I'm so glad fish don't have exoskeletons.

Sea turtle: Sorry. Not to say that you both wouldn't be delicious. It's just that I might choke on you. Look, here comes the shark. It might have an interest in eating you guys.

Fish: I'm not going to stick around and find out (exit).

Ammonite: Archie, can I hide behind your shell?

Sea turtle: No. Get away from me pesky invertebrate.

Shark: Hi Archelon. How are things in this part of the open ocean today?
Sea turtle: It's a little barren. There seems to be a shortage of jellies and I'm getting a little hungry.
Shark: I share your sentiment. I haven't eaten for a while and fish always swim away from me. Here, your fins look fleshy. Are you sure you need all four of those? (CRUNCH)
Sea turtle: (withdraws to shell, then smacks shark with flipper) Hey, knock it off you mindless eating machine.
Shark: Oh come on, you know me, I have my instincts. I just need to bite into some flesh right now. Just one bite.
Ammonite: Bite this. (inks in shark's face) I'm out of here. (exit)
Sea turtle: Me too. (exit)
Shark: What was that? Where did everybody go?
Elasmosaur: Come on little fishies, where are you? Come on little fishies, I like the way you move, I like the way you swim, I like the way you groove. Swimming in the current, either north or south, I want you to swim right into my mouth. Come on little fishies, listen to the song. Come on little fishies, play along. Oh what's this? It's the shark. You are a little too big for me buddy.
Shark: Elasmosaur, there aren't any fish around here and even if there were I wouldn't share them with you.
Elasmosaur: Come on Sharkie, do you really think your ancient methods can compete with my modern age fishing techniques?
Shark: You might think trolling through the water and snatching fish from the water works now, but believe me it will never last. My kind has seen creatures like you come and go over the ages. We sharks have been fishing for much longer than you reptiles have.
Elasmosaur: If that is the case then why are you getting thinner?
Shark: Hey, you gotta stay lean to stay keen. That's what I always say. Besides, it's good to have a little space in the open waters once in a while. But you can get out of here. I don't need you scaring all the fish away.
Elasmosaur: Hey. I go wherever I want, whenever I want and I don't need you to try and tell me otherwise.
Shark: Why you! This ocean's not big enough for the both of us. I'm gonna bite you!

Elasmosaur: You try that I'm gonna poke your eyes out.

Mosasaur: Ha ha ha, what's this all about? A long-necked euryaspid and a big-mouthed gill breather arguing over who rules the open ocean. Why don't I just swallow you both so you can discuss that in my stomach?

Elasmosaur: Oh hi Mosasaur.

Mosasaur: Who will be my prey today?

Elasmosaur: Oh Mosasaur, you might be big but you are too slow for me. You're just an overgrown lizard and a disgrace to other marine reptiles.

Mosasaur: (SNAP)

Elasmosaur: Yipes! (exit)

Shark: Oh gosh, I'm so hungry. I just have to bite into something right now.

Mosasaur: Ha ha ha. What are you doing, shark? What an amazing display of courageous stupidity. Time to crush your cartilaginous body. (SNAP)

Shark: Whoah! Okay, I'm leaving now. I'm so hungry. (exit)

Mosasaur: Ha ha ha, everyone knows that I am the pinnacle of marine animal evolution. Nothing can stop me. I am invincible. The oceans will be mine to rule as I see fit for the ages to come.

Act 1, Scene 4

Setting: An evening in the swamp. The swamp creatures wade.
Enter: Crocodile, lizard, frog

Lizard: Ah, this is nice. The water is cool and murky. Just the way I like it.
Frog: How was your day, lizard?
Lizard: It was just great. I creeped around, spent some time in the shade, ate some insects. All I could ever ask for really.
Frog: Yeah, this is the life. Amphibian heaven, I tell you.
Lizard: It's more like a reptile nirvana every day.
Crocodile: Oh hi fellas, mind if I join you.
Frog: Ahhh! There's a swimming dinosaur with sharp teeth in the swamp!
Crocodile: Relax fellow tetrapods. I am not a dinosaur, although I do not mind eating them. You, on the other hand, are hardly worth my time chasing after and digesting.
Lizard: Oh, you're the crocodile. You actually help us out by keeping big dinosaurs out of this swamp.
Crocodile: Thank you small reptile. I do my best. The dinosaurs may rule most of the Earth, but they don't rule the best part of it: the dirty, stinky swamps.
Frog: You got that right, ribbit.
Lizard: Why would we want to be anywhere else?

Act 2, Scene 1
Setting: A hot day on the plains. Green descends from the mountains, meets some creatures and delivers a message to them.
Enter: Green, Pterosaur, Pachycephalosaur, Ankylosaur, Ornithomimid, Tyrannosaur, mammal, bird

Green: I have been sent on a mission to restore peace and harmony to this planet and I shall not quit until my work is complete. Onward I descend into the vegetated plain, where my children the flowers show off their pretty colors, sweet aromas and fleshy fruits.

Pterosaur: Hark, who goes there?

Green: It is I, the green goddess. Mother of flowers and tender of all that grows in the ground.

Pterosaur: You are neither a dinosaur or a reptile. You're not covered in scales or fur. You don't look suited for this landscape at all. I mean I could eat you but I don't know how you would taste, you might be poisonous.

Green: Hah! I am not poisonous. I'm quite edible but I'm not to be eaten by animals like you today. I come here bearing a message for those who will hear it.

Pterosaur: What could you possibly offer me? I am the ruler of the skies. Your plants are useless to me. You are too colourful for my tastes but perhaps you can amuse me with your words.

Green: I come on behalf of the Earth to say that things are about to change dramatically here. Rocks will fall from the sky and there will be fire and brimstone and dust clouds. The very air will choke you and you shall be ruler of the skies no more, unless you listen to what I say.

Pterosaur: Ha ha ha, you do not scare me. I will not submit to you 'plant goddess'. It is you who should submit to me. Nature does not rule me. I rule nature.

Green: If you will not listen to the Earth then you will perish. For this planet cannot bear your strain any longer. I am here to provide shelter for those who are worthy. In time your fate will be revealed to you Pterosaur.

Pterosaur: Indeed it shall. I'll tell you how the story goes. I will continue to grow bigger, stronger and faster so that my shadow will block the light. Your precious flowers will wilt and I will feast on the carcasses of all the animals that once fed on them. That is my fate princess and I will decide it, not you. (exit)

Green: We shall see.

Pachycephalosaur: This is a creature I've never seen or heard of before. Is it an alien? Or an angel, perhaps?

Green: Good day dinosaur. My servants have revealed to me that you like to eat plants and use your skull to butt things. I would like to talk to you about that.

Pachycephalosaur: Tell me what you are first and what you're doing here.

Green: I am a guardian of the Earth and of all green things. I have come to the surface bearing a message that is both a warning and a promise. This world is about to change drastically. You must give up your material bulk and follow my footsteps to save yourself from the coming catastrophe.

Pachycephalosaur: That's pretty poetry, really, but I don't think your message is making it through this thick skull of mine. Look at it, it's so superior and intricate. Why would I ever want to give it up. Nature provided me with my skull and my size so that I could survive in this crazy world. If I were to give it up, I would become as useless as you. I think you are the one who should be worried about survival here. Watch your back sister. It's a dinosaur-eat-dinosaur world out there and only the strong survive.

Green: The time will come bonehead. Head to the cave in the north of the forest and prove your worth. Remember my advice and my flowers will be with you.

Pachycephalosaur: Yeah sure, your flowers will be with me. Get out of here.

Green: What has become of this planet? Has it been completely overrun with selfish ignoramuses that think my creations are only here to serve them indefinitely, just to fuel their pride and self gratification.

Ankylosaur: Doo doo doo, gotta eat lots of them to keep my armor strong. Gotta stay ahead of the game, that's what I always say. Gotta walk loudly and carry a big club.

Green: Hello.

Ankylosaur: What kind of creature are you? No sharp teeth, no tail, no horns, no spikes. Not much of a threat to anybody, are you? Are you lost or are you just here to offer your body to the carnivores?

Green: You are correct in saying that I am not a threat to anybody. In fact, I helped to create the flowering plants that you are eating right now. I am actually here to offer you advice.

Ankylosaur: Well in that case thank you. Although I do not see what advice you could possibly offer me. What is with that funny fur on your head?

Green: I am a servant of the Earth and I come bearing a message. The world is strained and its clock needs to be reset. I am here to offer shelter for animals who are worthy before the storm comes. If you leave your unnecessary bulk and extremities behind and follow me to the forest I shall protect you.

Ankylosaur: Whoah now! Hold on! I am the Ankylosaur. Do I look like I need protection? This armor has been built up over millions of years and you want me to give it up? Not even the fiercest of dinosaurs can hurt me. I think it is I who should be offering protection to you, flower lady.

Green: You must look beyond the immediate my friend and into the bigger picture. There is much more to this planet than large beasts. The Earth is the home of many creatures big and small, old and new, two-legged, four-legged and no-legged.

Ornithomimid: Ankylosaur! The Tyrannosaur is coming.

Ankylosaur: You just go tell that tyrant lizard that I'll be right here waiting to crush some bones. If he messes with me, he won't be getting away with just a sore toe today.

Ornithomimid: Who is the creature standing beside you?

Ankylosaur: I don't know, some hippie lady, selling karma or something.

Green: I am a servant of the Earth and I come to the surface bearing a message that must be heard.

Ornithomimid: Listen, I don't really have a lot of time to talk right now. We should all get out of here soon so that we don't turn into carcasses.

Pterosaur: Ornithomimid. Just the dinosaur I have been looking for.

Ornithomimid: Ahhh! Pterasaur! Gotta run.

Pterosaur: Don't do that.

Ornithomimid: Into the forest. The Pterasaur cannot catch me there. (exit)

Pterosaur: You may be swift and agile but you cannot hide forever, I will get you next time.

Bird: What happened? Did I miss the fight?

Ankylosaur: No, you are just in time.

Tyrannosaur: Ankylo! We have unfinished business. Let's settle this once and for all time.

Ankylosaur: Come and get it tyrant lizard. My tail's ready to bust some ankles.

Tyrannosaur: RAAR (Tyrannosaur advances, Ankylosaur swings, Tyrannosaur dodges, Tyrannosaur bites Ankylosaur's shell, Ankylosaur breaks free, Tyrannosaur advances again and bites, Ankylosaur dodges, Tyrannosaur steps on Ankylosaur's tail)

Ankylosaur: Ow!

Tyrannosaur: You are not so tough without your tail. Now let's see how your leg tastes.

Ankylosaur: AHHH (Ankylosaur rolls on top of Tyrannosaur's head, Tyrannosaur loses balance, Ankylosaur breaks free, manages to get up again and crawls away with its tail dragging behind on the ground and bleeding, Tyrannosaur stumbles but regains balance)

Tyrannosaur: It is over for you Ankylosaur. You won't get away now.

Mammal: (yawn) Where is all this noise coming from? How am I supposed to sleep with all this racket?

Tyrannosaur: I'm coming for you Ankylosaur. Enjoy your last moments of locomotion before I tear your flesh apart.

Green: Oh no, that furry little creature is going to get trampled. I'll going to put an end to this ridiculous competition before something else gets killed unnecessarily.

Bird: Ankylosaur! Use your tail! That always works.

Ankylosaur: I can't use it anymore. I think it's broken.

Bird: Just pretend it's still working. Rex is afraid of it.

Ankylosaur: I can't tweeter. I have to run to safety.

Tyrannosaur: You have met your match today armoured lizard.

Green: Stop right where you are dinosaur! I need to talk to you for a few moments.

Tyrannosaur: Who are you? Are you offering yourself to save the Ankylosaur? Just stand aside and I will eat you later. What happens between me and Ankylosaur now is personal.

Green: Actually, I came to warn you all that a great catastrophe is coming and even great beasts like you will not be able to survive. You must heed my words carefully. Stop this ridiculous behaviour, cast off your material bulk, take only what you need and follow me into the caves of the Earth or be consumed in the flames, ash and death that await this place.

Tyrannosaur: RAAAR! I am the top of the food chain. Nature bends to my will. I will eat you and then finish off the Ankylosaur. Prepare yourself.

Green: What an arrogant beast. I will have to use my powers to protect myself. Green things of the earth, rise up against the tyrants of the world. (Green throws seeds at Tyrannosaur, they quickly grow to vines that tie Tyrannosaur to the ground, Tyrannoaur struggles for a bit and then breaks free)

Tyrannosaur: Hah! You can not stop me with your magic. What good are plants against ten tons of meat and razor sharp claws and teeth?

Bird: Hey mother nature, use the trees with needles. Rex doesn't like forests. He can't run though the big tree trunks.

Green: I'll see what I can do. (Green throws pine cones on the ground that grow into big trees quickly, Tyrannosaur stops for a moment and then knocks one of them down with his tail which comes to the ground with a thud)

Mammal: Eep!

Tyrannosaur: It is no use. I am just too powerful. Just surrender your body and have me do what I will with it.

Bird: Uhh, mother nature, just in case you don't survive this, I wanted to let you know that your creations are really beautiful. Especially your flowers.

Green: That's it. My greatest creation. How could I forget?

Tyrannosaur: This is the end for you hippie princess.

Green: You may be able to break through vines and knock down trees tyrant lizard, but you are no match for the power of the flower. (pulls out flower)

Tyrannosaur: Ha ha. What?

Green: I believe in you my children. (flower is swarmed by thousands of insects, Green throws it into the mouth of Tyrannosaur)

Tyrannosaur: (CRUNCH)(eats flower, insects turn on Tyrannosaur, fly through his nose and his mouth, into its eyes, biting and stinging) Ow! That hurts. What magic is this? Get them off of me, get them away, Ahhh! (exit)

Green: Phew! Well I'm happy that worked.

Mammal: That was incredible. You scared away the Tyrannosaur with a flower.

Ankylosaur: Ah, my tail!

Green: Ankylosaur, you're bleeding. Here let me help you. The power of my plants can heal you.

Ankylosaur: No, let me be. I must toughen up for the next battle.

Green: Where I'm going you won't need armor or weapons. Just listen to me. Give them up and follow me into the forest.

Ankylosaur: No, I cannot, you may have saved me today but my place is here. My plates and my tail are a part of me. I cannot give them up.

Green: Sure you can. If you could only see the prize that awaits in the next world for those who are willing to make the sacrifice.

Ankylosaur: If my world is to die, then I shall die with it as a warrior in their kingdom. Best of luck to you. Watch your back. You don't have armor like me.

Bird: I will come with you Mother Nature. You are my hero. I shall follow you wherever you go.

Green: Let us leave this forsaken place and go to the forest where the rest of the creatures await me.

Mammal: Lead on green goddess. I will follow the trail of insects.

Act 2, Scene 2

Setting: Mid-afternoon in the forest. Green meets the herbivores and delivers a message.

Enter: Green, Hadrosaur, Ornithomimid, Dromaeosaur, Ceratopsian, mammal, bird, snake

Ornithomimid: There might not be as much room to run around in the forest, but the shade sure is nice.

Green: Look at these trees. My, they certainly have come a long way.

Ornithomimid: Oh it's you again. Sorry I had to run last time but you know how it is with those carnivores. Running is what I do and nobody does it better than me. Look at me I'm so shifty and nimble.

Green: Spare me that monologue. I have had big enough dose of dinosaur egos for one day. You don't have to tell me how great you are, for where I am going skills like that won't really matter.

Ornithomimid: Where are you going?

Mammal: We are going to the cave and to the refugium on the other side of the underground passage.

Green: Well yes, but we will have to set it up first.

Ornithomimid: But that cave is dark and uninviting. Only the vilest of creatures dwell there.

Hadrosaur: Hi flower goddess. The bird told me that you defeated Tyrannosaur in single combat today. Even the Ankylosaur and Ceratopsian can find that to be a real challenge.

Mammal: It was great, she threw a special flower at Rex and all the insects from miles around swarmed the tyrant and chased it away.

Ceratopsian: It's about time someone gave that braggart a whooping. Who are you anyway?

Green: I am the green goddess. I created this forest long ago. I am a servant of the Earth and I have come to the surface with a message for those who will hear it.

Hadrosaur: Sure, tell us what's on your mind?

Green: The Earth tells me that it is strained and weary. It tells me that the large, bulky dinosaurs take too much and return little. Soon there will be a catastrophic event to reset the system. All of the Earth's creatures, big and small, must follow me or be consumed in a fiery inferno.

Ceratopsian: And to where would we be following you?

Green: I am going to prepare a refuge, through the cave and into the mountains where we will be safe. You'll have to give up your material bulk and extravagant accessories. They are a burden to you and you will not need them where we are going.

Ceratopsian: You mean my headcrest and my horns? Those are my greatest attributes. Do you mean to tell me that you want me to look like a wimp?

Green: It is both a warning and a promise. Those who are willing to make this sacrifice will inherit the new world.

Ceratopsian: But I can't give up my frill and my bulk. That would make me vulnerable to attack like my ancestors were. This frill represents millions of years of hard work and patient evolution. The animals that don't follow your advice will simply take advantage of the ones that do.

Hadrosaur: And princess, what would I to do? I can't give up my bill because it helps me eat lots of plants to sustain my large body size.

Green: Then you must fast. Eat less so that your body will shrink and it will not have energy to expend on accessories. Go back to the basics I tell you.

Ceratopsian: You are such a hippocrate. Look at all that extra space you have in your skull. And what about those fancy digits on your front paws? And what are those other appendages for?

Ornithomimid: What trickery is this?

Green: Hey, I am not here to stay. I was just sent here on a mission to save this planet that I love.

Ceratopsian: Look, my kind has made it through many challenges and I'll get through whatever comes my way. I don't need your help or advice.

Green: Please Ceratops you must understand, I want the best for all creatures big and small. But the current state of the world cannot be sustained for much longer.

Snake: I sssense great danger. I know the underground passage that you speak of. It does not lead to a promised land. It is dark and narrow and only the nimblest of creatures, such as I, can navigate through it.

Green: If you listen to my words I will clear the way for you to travel through it.
Snake: By all means, travel the dark road. I care not either way.
Dromaeosaur: My friends, do not listen to this creature for it is a trickster. It wants you all dead and for your children to become its slaves. There is no promised land. It is leading you to your graves. You will starve and never see the light of day again.
Hadrosaur: You little thief. How dare you show your face in this forest again Dromaeosaur. I should stomp you into the ground.
Ornithomimid: You know, it may take a trickster to know one, Hadrosaur. Dromaeosaur may be right. What this plant goddess says doesn't sound right. If the grass really is greener on the other side of the underground passage, then why wouldn't the snake have stayed on the other side.
Ceratopsian: Yeah!
Hadrosaur: Yeah!
Mammal: Because snakes don't eat grass?
Ornithomimid: It's just a metaphor.
Green: Please, noble creatures you must trust me. If you come with me, then together we can build a refuge which will be the foundations of a better world. It is the only way you can survive.
Ceratopsian: I don't believe you. If you really want us to do what you say then let's see you give up your extra weight.
Green: Please creatures. It is not my choice to make. You must either take my words to heart or be destroyed by powerful forces from outside that even dinosaurs stand no chance against.
Mammal: I believe you, plant lady. I saw you turn away the Tyrannosaur. If you can do that then you can do anything.
Bird: Lead us to the refuge, wherever it is, Mother Nature.
Green: Thank you small creatures.
Snake: I couldn't care lessss.
Ceratopsian: I have heard enough of this nonsense. Enjoy your last moments of sunshine. You go to you deaths, little beasts (exit).
Ornithomimid: Hold on! I am willing to give you a chance to prove yourself. If you go through the cave and don't die then I might consider your words.

Hadrosaur: Let me know how the experiment goes. I'm just going to digest some more plants (exit).

Dromaeosaur: You go to your deaths little ones. She is seducing you with her charms and fantasies. Come this way. I shall protect you from her evil spells.

Green: No raptor, they are free to decide their own fate and so are you. If you choose to act selfishly and dishonestly, then you will soon see the retribution for your actions. Come now my noble creatures, to the cave. The future of this world awaits us.

Act 2, Scene 3
Setting: A sunny day on the coast. Blue enters the water, meets sea creatures and delivers a message.
Enter: Blue, lobster, ammonite, fish, Shark

Blue: I love the beach. The feeling of a coastal breeze on my skin, the smell of the salty surf. The water invites me. Let's see what colourful treasures await me in the shallow waters of the world.
Lobster: Hi there!
Blue: Greetings my shelled friend. I am siren. I am here on a mission on behalf of the earth. Can you tell me what you do here?
Lobster: I am the lobster. I sort of act as a groundskeeper here in the reef. I trim the lawn and keep all the creatures in the benthos in check.
Blue: I appreciate your efforts, but I notice that there are some creatures that you are unable to control.
Lobster: I suppose I do my best with the limitations I have.
Blue: Now here is an impressive creature. Look at that beautiful shell.
Lobster: Oh no siren. That is the ammonite, a proud and stubborn beast. It is not my friend.
Blue: What grievances do you have with this creature?
Lobster: You see miss. Long ago its kind stole the sea from my kind, the arthropods. In the Cambrian period we ruled the shallow seas. Our shelled bodies and jointed appendages performed a great variety of specialized tasks. Everything was perfect. But then the cephalopods came around. Ruled by greed and decadence, they exploited the seas and one-by-one took valuable niches away from the arthropods. Eventually a new group, the vertebrates, superseded us both but one day I would like to see the glory of the arthropods restored.
Ammonite: Stay in the dirt crustacean. You are an eyesore of a creature.
Lobster: Cephalopod scum!
Blue: No need to fight you two. I am here to deliver a message for all the creatures of the sea to hear.
Ammonite: Who are you?
Blue: I am the siren.

Ammonite: What is your message?

Blue: The time has come, the earth will shake. Meteorites, tidal waves, volcanoes and earthquakes. The ocean will become red with blood. Millions of bodies will sink to the mud. Come with me now if you want to survive. Leave all you don't need and take only your lives. Follow me now before the ocean gets odder. Follow me now upstream to freshwater.

Ammonite: You seem to be in good spirits for someone who is prophesizing doom? Why should I listen to you? What power do you have?

Fish: Hey guys, what's going on?

Ammonite: This crazy creature claims that the world is going to end and we are all going to die unless we follow her into a river.

Blue: The ammonite is correct. I'm on a mission to preserve the seeds of life in the ocean for the coming Apogalacticon.

Fish: Do you have any songs that are a little less depressing?

Shark: I have a new song that's kind of catchy. Crunch, Crunch, Crunch, Crunch. I want you in my lunch. I want to have some dinner before I get thinner. Crunch, Crunch, Crunch, Crunch. Fish and squid for lunch. Let's spend the day together, together in my gut.

Fish: Ahhh! The shark is here. (hides behind corals)

Ammonite: Eek! (inks in water)

Shark: (cough) Where'd you go? Come back here.

Blue: Hey teeth! Why don't you pick on someone your own size?

Shark: Who are you and what do you taste like?

Blue: I am the siren, the messenger of Apogalacticon and the preserver of life.

Shark: Nice to meet you. I am the shark and I am hungry.

Blue: Listen shark, you should consider changing your ways if you want to get through the coming catastrophe. You are going to need to adapt. You might need to fast and survive in freshwater for a little while to get through this turbulent time.

Shark: But I'm hungry. Why don't you just let me eat these small animals?

Blue: I need to find some animals to populate the post-apocalyptic world. I want to give them a chance to prove themselves.

Shark: Fine then, I will eat you instead.

(Blue dodges behind corals, shark bites corals, blue dodges behind giant clam, shark eats giant clam, blue stabs shark, shark cries)
Shark: WAAHHH! I just wanted a little taste. Oh look, something shiny!
Fish: Ahhh! (exit).
Shark: Fishy! (exit)
Blue: It is time for me to spread my message to deeper waters. (exit)
Ammonite: This is gonna be good. (exit)

Act 2, Scene 4

Setting: The open ocean, Blue goes to the open ocean and talks to creatures.

Enter: Blue, ammonite, fish, Sea turtle, Elasmosaur, Mosasaur

Fish: I don't know how much longer I can swim with the shark chasing me like this.
Shark: You are mine, fishy.
Sea turtle: Oh no, not you again.
Fish: Hey Sea turtle, can I hide behind your shell?
Sea turtle: Oh no, please just get away from me.
Shark: Where is the fish?
Sea turtle: It's behind me shark. Don't bite me.
Elasmosaur: Shark, how dare you show your mouth around here again.
Shark: Oh not this again. I hate these long necked reptiles.
Elasmosaur: I told you. This is my ocean.
Fish: Everyone wants to eat me.
Blue: Come fish. This way. To the surface. I will protect you.
Elasmosaur: Who are you and what are you doing in my ocean?
Blue: I came to deliver a message. Hear it now loud and clear. Prepare yourselves because the world is going to change and there is going to be a big crisis. The clock of the Earth is going to be reset and every animal in the open ocean is going to be tested. You must fast and give up your carnivorous ways for a little while. Follow me to the mouth of the river if you want to survive.
Shark: Hey Elasmosaur, maybe we could each eat half of this two-legged thing.
Elasmosaur: I think I can just sneak right around this swimmer lady and retrieve the fish. Watch this.
Blue: Just stay close, fish. I'll take care of this monster.
Fish: Eek! I don't want to die. (Elasmosaur moves neck around and snatches at fish, but blue twirls and stabs Elasmosaur in its throat)
Elasmosaur: Ouch!
Mosasaur: BURPPP!
Sea turtle: I'm getting out of here. Bye guys. (exit).
Shark: I am suddenly reminded of a biting wound I received recently (exit).

Elasmosaur: Nice meeting you. I will chase that fish again another day (exit).
Blue: Both of you have wounds from my dagger to remind you of the coming catastrophe. Remember my words for when the time comes you will be judged.
Mosasaur: Who ventures into my domain?
Blue: It is I, the siren, here to warn you of the coming Apogalacticon.
Mosasaur: Make it quick for I must feast.
Blue: Arrogant and gluttonous beasts, such as yourself, will be too complacent in their habitats, making them unable to adapt to environmental stresses.
Mosasaur: I am the Mosasaur. I'm a big, tough reptile and I can handle whatever stress the environment throws at me. The question is, can you handle the stress I am going to place on your environment?
Blue: You have a long way to go Mosasaur but if you listen to my words, I might be able to help you.
Mosasaur: I think I might be able to help you too, but first let me swallow you.
Fish: What are we going to do? He's going to eat us.
Blue: Just stay close fishy. This monster will not prevail.
Fish: He's coming..
(Blue dodges Mosasaur's bite, twirling and slashing at Mosasaur with her dagger)
Blue: It's no use fishy, this beast is too big. My dagger cannot puncture its skin.
Mosasaur: It's over for you, siren lady.
(Blue dodges Mosasaur again, and grabs a hold of its hind flipper, Mosasaur tries to shake her off)
Mosasaur: Get off of me!
Blue: This beast is too powerful. We have to get to shallower water, fish. Which way to the reef?
Fish: That way.
Ammonite: Ahhh! Mosasaur! (struggles to move out of the way)
Blue: Meet me at the reef fishy.
Fish: Okay.
(Blue lets go of the Mosasaur and grabs the ammonite)
Ammonite: Hey, let go of me!

Blue: Ink, ammonite. Ink like you've never inked before!

Mosasaur: Time to meet your end.

Ammonite: Ahhh! (Ammonite releases a cloud of ink. Blue dodges the Mosasaur and drags the Ammonite behind her as she swims around the Mosasaur in a circle and then exits in the direction of the reef.

Mosasaur: (cough, cough)

Act 3, Scene 1

Setting: The underground passage, green enters cave and summons insects.

Enter: Green, mammal, bird, snake, Ornithomimus, Dromaeosaur, spider

Snake: Here we are. Thisss is the cave.
Bird: It looks so dark and scary in there.
Mammal: It seems kind of cozy.
Dromaeosaur: You'll find nothing but death in there, I tell you.
Green: This will be challenging. I must call upon my army of followers. Insects from all parts of the forest, I call upon you to help us find our way through the underground passage and lead us to the promised land.
Ornithomimid: I won't be going through there this time (exit).
Dromaeosaur: Ick! Bugs everywhere. I'm getting out of here (exit).
Green: Snake, can you see anything in here?
Snake: I don't need eyes here, I use my tongue to sssenssse my sssurroundingsss.
Mammal: I can't see anything. How do we know if this place even leads anywhere?
Green: Light bugs to the front please (fireflies go to the front and light the way, but then they get caught in a spider web).
Spider: Well hello there, what a nice surprise.
Snake: Ah, you again. I thought you might be here.
Spider: Good to see you snake, thanks for bringing some tasty friends with you.
Green: Excuse me octopede, but you can't eat those. We need them to find our way through this tunnel.
Spider: These bugs are stuck in my web. They are useless to you now. Let me eat them.
Green: If you come any closer to them, I will destroy your web and you with it.
Spider: Please, I am a simple creature. I eat only what lands in my living room.

Mammal: Pardon me simple creature but we are searching for a way through the cave. Do you know how to get through to the other side?
Spider: You want to get to the other side? There is nothing there.
Green: We are going to build a refuge in the mountains. It will be a safe place to preserve the seeds of life through the coming apocalypse.
Spider: I will not let you through unless you pay me a tax.
Green: Don't be selfish. If you abandon your web and leave everything you don't need behind, you can come with us.
Spider: I can let you through the passage after you satisfy my hunger.
Green: That is what you want? My insects?
Mammal: My insects!
Snake: My insectsss!
Spider: Yes, what do you think I built this web for? Insects are what I crave, like this lightning beetle right here. (eats insect)
Green: That's enough. Let us through the passage or I will squash you.
Spider: No, step back, one bite from me and you will be poisoned and frozen in your tracks forever. The way is shut. It was made by the webmaster, and the webmaster keeps it. Give me all of your insects to satisfy my hunger and I will let you through my cave in peace.
Mammal: You selfish beast, we need those bugs to survive.
Bird: It is unwise, arachnid, to take them all at once.
Spider: Silence dinosaur, I will keep them in my web to feed myself for an eternity, you can eat something else from now on.
Green: Alright, calm down. I understand your demands spider. I cannot give you all of the insects but I'm sure we can compromise. If you give me back my light bugs, I will give you one of each kind of insect. The rest will survive to restore the population so that you will be able to eat more later.
Mammal: Aw, I want to eat insects now too.
Green: No my little helpers, you must fast. The insects that are with me now will be the ancestors of the insects in the next age. Now, webmaster, tell me where the passage to the other side is.
Spider: Okay lady, here are your light bugs. Put the other insects you promised in my web. If you don't bring me more later, then I will come to collect them with a vengeance.

Green: (summons a variety of insects and places them in the web) Alright, here they are. Sorry little ones but your survival will have to come with some sacrifices. Now let us through or I'll tear your web apart.

Spider: Relax lady, let me fold the web gently so that my precious food will be stored. I will build another one after. The passage continues through my web. Just keep going, turn around the corner, and follow the light to the surface. I don't really like it over there because it's too sunny.

Green: Okay creatures follow me. We have work to do. (exit)

Act 3, Scene 2
Setting: A pristine freshwater lake. Green and her followers emerge from the underground passage and starts creating a refuge.
Enter: Green, mammal, bird, snake

Green: Oh, it is so great to feel the rays of the sun again.
Mammal: Wow! A waterfall and a pond.
Bird: This place is very isolated.
Snake: I couldn't care less! I preferred it in the underground.
Green: This will be the refugium. Okay insect servants. Those of you that fly, go forth and retrieve the seeds of your favourite plants and bury them in the fresh sediment by the pond. Those that crawl, build walls of soil and bunkers for our vertebrate friends to hide in.
Bird: What are we to do, Mother Nature?
Mammal: I'm going to have a feast with all these insects and seeds.
Green: I have other tasks for the three of you. For now you must fast. You can eat when the battle is over.
Mammal: Battle?
Snake: I did not sssign up for thisss!
Green: Bird, I will need you to help clean the skies. Mammal, you will help relieve the land of the undeserving. And snake, I may need you for another task. Go back to the plain and report to the blood maiden whose name is Apogalacticon, the destroyer of life. She will make an impact when she comes. This will be followed by dust clouds, tidal waves and volcanoes. All creatures on the plain and in the forest will be killed except those creatures with my seal on their foreheads.
Bird: You mean all my cousins are going to die? (Green marks bird)
Mammal: It sounds like it's going to get dangerous out there. (Green marks mammal)
Snake: Why would she kill all the other creatures? (Green marks snake)
Green: They have ignored my advice and have not shown their worth. They are too large, they eat too much and so they cannot adapt. They won't be able to crawl through the underground passage. But the bloodlines of the three of you will be enough for vertebrate life on this planet to recover. Now go to the plain. Your destiny awaits.
Bird: Yes, Mother Nature. It shall be as you say. (exit)

Mammal: This better be worth it. I'm getting kind of hungry (exit).
Snake: I don't want to go. I want to ssstay.
Green: You have to go snakey. I need some time alone with the new forest.
Snake: Sss, why can't I remain in this place?
Green: Because I will beat you with this stick if you don't leave. Go now snake and prove your worth to me.
Snake: Alright, alright, I'm going. (exit)
Green: Tall trees and bushes will be needed to protect this place from the terror that awaits the rest of the world. Thank you my insect friends. Keep up the good work. We have a long way to go.

Act 3, Scene 3
Setting: A hot day on the coast. Blue sorts through the creatures of the reef. Fish kills Elasmosaur.
Enter: Blue, lobster, ammonite, fish, Sea turtle, Shark, Elasmosaur, Mosasaur

Blue: I'm glad that worked, we should be a little safer here. That monster won't be able to manouver in the shallow water. Ok followers, we have some important things to take care of. I will need your help.
Fish: What do you need me to do?
Blue: For now, please keep a look out. Guard the reef while I sort out the reef invertebrates.
Fish: Oh, I don't want to deal with predators.
Blue: Oh, but you must. I require this task of you. It's a dangerous ocean out there fishy so take my dagger. It is called Teleos. It will help you swim faster and help ward off predators. It will serve you well.
Fish: Oh, I'm not much of a fighter.
Blue: So be it, I knight you sir Fishy of the ocean. Follow me you two.
Fish: I'll do my best.
Blue: I need to select the seeds of life among the organisms of the reef to be preserved through the coming storm.
Ammonite: You have caused so much trouble already, just go away.
Blue: I will be gone soon enough. You can come with me, or do you want to stay here with the predators?
Ammonite: I just want to be left in peace.
Blue: Swim to the freshwater lake then. Although you will have to give up your shell in order to swim through the river current.
Ammonite: An ammonite cannot be without its shell. It is my best attribute.
Blue: We don't have time for this argument. I'm going to just add your shell to my jewelry collection. (takes shell).
Ammonite: Ahhh, I'm naked.
Blue: Lobster, there you are. We have to sort through the reef creatures and select the seeds of survival. Let's take some sponges,

and some gastropods, and some bivalves. But not those heavy rudist clams, we'll just leave them behind.

Lobster: Hey I was going to eat that one.

Blue: Oh no lobster. You will have to fast for a while as we make the journey to the refuge. Okay, let's take some astroids and echinoids and holothuroids.

Lobster: What? I'm not going anywhere.

Blue: Do you want to stay here and die instead? Corals, anemones, crinoids. It is so much easier with these sessile epifauna. All I have to do is squeeze them into a container of salt water and carry them up the river.

Lobster: Where are you taking all that epifauna?

Blue: To the refugium of course. Come on you too, hurry up.

Fish: Siren, there is a predator coming this way.

Blue: Which one?

Fish: Eep!

Elasmosaur: There you are.

Blue: Are you ready for the journey to the lake, Elasmosaur?

Elasmosaur: No, but I am ready to eat the fish.

Blue: I don't have time for this right now.

Fish: What am I supposed to do? He wants to eat me.

Blue: Just use the dagger and keep it close. I believe in you fishy. You can outswim and outsmart this brute. Just remember where you must meet me after.

Fish: Uh, stand guard you long-necked fiend.

Elasmosaur: Just stay right where you are fish. My mouth will be there very soon (snatches at fish).

Fish: Ahhh! (swallowed whole by Elasmosaur)

Elasmosaur: (gulp)

Ammonite: Poor brave vertebrate. At least it was able to become a good meal.

Elasmosaur: Ha ha ha, I'll eat you next.

Blue: You've already swallowed more than you can chew today, Elasmosaur.

Fish: (cuts open Elasmosaur neck with blade and swims out) You won't be eating anything ever again, long neck!

Ammonite: Whoah, I thought you were dead for sure.

Lobster: Man, all those bones and meat could feed me for months.

Blue: And here, we'll take some brachiopods and bryozoans to add a little variety to the ecosystem.

Fish: I did it. I defeated the Elasmosaur. And I did it all because of you Siren.

Blue: Good work fishy. I will let you keep Teleos. You can wear it as your new fin to help your kind in the future.

Fish: Oh thank you so much Siren. How will I ever repay you?

Shark: Oh hello again. I smelled blood around here. Oh look, it's the fish and the squid. There will be a party in my mouth and they're all invited.

Blue: Stop right there shark. The blood you smell is from the corpse of Elasmosaur, over there. However, I think you are a survivor. You should fast a little bit longer and head up the river with us.

Shark: Nah, I just need to eat something right now. (bites Elasmosaur body) Take that Elasmosaur. Do you like it when I bite your ugly neck like that, huh? Not so tough now are you?

Blue: I will need a crustacean, they are good at surviving. Come on lobster, come with me up the stream, we have to get to our safe spot. Sharkie, come on this is your last chance. Stop eating right now and follow me to the stream or there won't be any sharks in the ocean ever again.

Shark: Mmm. Eating an Elasmosaur is like eating a thousand fishies all at once.

Mosasaur: (CRUNCH)

Shark: The neck was delicious, but where did the body go? Mosasaur, what are you doing here? Give me back my main course.

Mosasaur: Back off Sharkie, this kill is mine.

Shark: No.

Mosasaur: I said back off!

Shark: Ahhh! Hold on Siren of doom, I'm coming with you. (exit)

Mosasaur: Ha ha, where are you going? Without you and Elasmosaur, I guess I'll just have all the prey in the ocean to myself.

Act 3, Scene 4

Setting: A hot day In the swamp. Sea turtle lays egg, Crocodile eats Hadrosaur.

Enter: Blue, lobster, Sea turtle, frog, lizard, Crocodile, Hadrosaur, Shark

Blue: Here it is, the mouth of the river.

Lobster: Ick! how am I ever going to make it upstream? The current is so strong.

Blue: Don't be a silly crustacean. You have big arms and pinchers. Use them to grab on to vegetation, like this. Hey, it's the Sea turtle. Archelon! Hey, we have to leave this place now. The great storm is coming. Come quickly to the refuge. Time is running out.

Sea turtle: I won't be able to propel my heavy body all the way up that river current. But I want to give you my egg. You can take it to the refuge. I want my offspring to live on.

Blue: You can trust me dear reptile. I will bring your egg with me to the freshwater lake.

Sea turtle: Farewell, my brave offspring. (gives egg)

Blue: Goodbye Archelon as we must go quickly. Onward my followers, we must continue our journey up the stream.

Frog: Ribbit.

Lizard: (slurp)

Blue: Hi there, little tetrapods.

Frog: Ahhh! It's a swimming dinosaur with hair on its head!

Blue: Relax amphibian. I am not a dinosaur, nor do I wish to eat you.

Lizard: What are you doing in the swamp, strange creature?

Blue: My followers and I just need to pass through this area on our way to the freshwater lake. We left the sea to escape a catastrophe that is swiftly approaching. If you want to survive you should come with us.

Lizard: But this is my home. I never want to leave this place.

Frog: It's our home. We can't just leave everything and go.

Blue: Then I'm sorry my friends, but you will go extinct just like all the other stubborn life forms on this planet.

Hadrosaur: This is great. I love coming to the swamp. Oh its another one of those weird creatures. What are you doing here?

Blue: I am here to bring all the water-dwelling animals who are worthy to a refuge that will preserve the seeds of life through the coming apocalypse. Are you prepared to leave what you don't need behind and follow me up the stream, dinosaur?

Hadrosaur: I have heard a similar message from the other strange creature dressed in green. It led some creatures into an underground passage and they never came back. I mean, there could very well be a refugium on the other side of the tunnel, but I'm just too big to make it through.

Blue: Well that's okay my duck-billed friend. You can follow us through the swamp. We are going to the same place.

Hadrosaur: Well, if what you and the other creature say is true, that there is a disaster coming and all life will perish, then I guess I have little choice. But can you offer me any proof that what you say is indeed the truth?

Blue: I cannot offer you any proof. You have to trust my words. Follow me now or be consumed in the fiery blaze that awaits this place.

Hadrosaur: Well, you could very well be right. I have seen a lot of strange things lately. I don't really have a lot to lose coming upstream with you.

Crocodile: (SNAP) Gotcha! (Crocodile eats Hadrosaur)

Hadrosaur: Curse you crocodilian!

Blue: Hey, what just killed one of my potential followers?

Lizard: That's the crocodile.

Frog: It likes to eat dinosaurs.

Crocodile: Oh hi there, I was planning to eat you but then I saw the Hadrosaur. My name's Deinosuchus. How do you do?

Blue: Thank you for not eating me. I am fine. I have visited the ocean and spread my message to all the creatures there. And now my journey has led me to this swamp. A big catastrophe is coming soon. We are heading to the freshwater lake where we will be safe. As long as you don't eat any more of us, you can follow me.

Crocodile: Once I stuff this Hadrosaur carcass into my belly I will have enough food to last a long time. Tell me about this catastrophe.

Blue: Our mission was to warn all creatures, including dinosaurs, about the coming Apogalacticon, a time of great destruction where forces from outside the Earth will have an impact and forces from inside the Earth will erupt. It will be a period of great change for the Earth's ecosystems. All of the animals that are deemed worthy by us who are willing to make the journey are being directed to a designated refuge where they will be sheltered from the destructive power of Shiva, the destroyer of life.

Lizard: Wow. That sounds interesting. Tell us more.

Blue: There are only two ways to get to this refuge: through the underground passage or through this swamp. These paths block out the destroyer's influence and the destroyer knows not to go past them. All creatures of the Earth who do not heed our advice and make their way to the refuge in the mountains will meet their end.

Crocodile: This refuge sounds kind of like a nice place. Have you invited any of the big dinosaurs?

Lizard: I hate dinosaurs.

Blue: I have not met any dinosaurs except the Hadrosaur you just ate.

Frog: I want to go to the lake.

Lizard: Me too. I can come to the refuge but this better be a good gathering, or else I'll heading back to the swamp.

Shark: I'm glad to escape the jaws of the Mosasaur once again, but I don't know how long I can stay alive in water that doesn't have salt.

Blue: Sharkie, you made it.

Shark: Siren, I want to come with you. I've had a meal but now the ocean is just too scary. Hey, what is that smell?

Lizard: You must be referring to the blood of the dinosaur the crocodile just killed?

Shark: I love that smell and all of a sudden I'm starting to feel a little hungry.

Blue: Shark, come on. Please control yourself.

Crocodile: Well, you can't eat my dinosaur.

Shark: I have to eat something.

Frog: Ahhh!

Lizard: You wouldn't like me, I taste kind of sour.

Blue: Quit fooling around sharkie. If you're coming with me you'll have to go on a special diet for a little while. Here, suck on this salty rock for a while. (puts rock into shark's mouth, shaped like soother)

Shark: Mmm.

Blue: Alright my brave followers, to the lake we go!

Act 4, Scene 1

Setting: An early morning on the coast. Red crashes into the shallow water, makes a crater and destroys the creatures of the reef.
Enter: Red, Sea turtle

(Red comes in with a meteor that makes a big boom and a splash, the ground is scarred with a crater)
Red: At last! Apogalacticon has arrived to reap the overripe fruits from the vines of the world. Here I stand a mighty amazon, a dragon slayer and destroyer of life. It shall be as prophesized. I will extinguish these pitiful life forms and watch the sea turn red with their blood. And I will start with you, creatures of Chicxulub. This shall be the end of your pathetic little reef. From now on you will be remembered only in the fossil record.
(Red slices up sponges, clams and corals and then hurls grenades around the reef that explode and spew forth black toxic smoke)
Red: Ha ha! Yes that´s right. Suffocate on these fumes. Feel the death spreading all around you and join in. The end of the world is here to erase all of your names from the book of life.
Sea turtle: Whoah! What's going on in the reef today.
Red: Halt creature! I have a message for you.
Sea turtle: The blue creature already told me the message. Who are you?
Red: I am Apogalacticon, the destroyer of life, here to fulfill the will of the galaxy, to relieve the strain of overpopulated ecosystems around the world. Do not struggle, your fate is to die here and now and I am to realize it.
Sea turtle: It is just as the siren foretold. I must get out of here.
Red: No, there will be no escape for you. You shall perish by the tip of my spear.
(Red stabs spear through turtle shell)
Sea turtle: Ahhh!
Red: Ah yes! So shall it be written in the stars and so shall it be done. Just like this reef, all other ecosystems will crumble and burn. That is all to come shortly in the wake of my armageddon.
(Tidal wave sent forth from the reef, sending ripples through the ocean and crashing down on the land)

Act 4, Scene 2

Setting: Midday in the desert. Red orders bird to destroy Pterasaur's egg. Pterasaur flees from Red but is eaten by Mosasaur.

Enter: Red, bird, Pterasaur, Ornithomimid, Mosasaur

Red: Beyond the decapitated reef lies the desert, where the hardiest of life forms have scraped out a living in desolation, until today of course. Burn, burn, plants of the desert. Whatever life that calls this place home will meet its end today.

(Red throws burning rocks that explode and ignite a set of blazing fires)

Ornithomimid: Ahhh! Fire! Smoke! Ash! Brimstone! The apocalypse has come.

Red: I've come to cleanse the land of overgrown dinosaur filth like you, Ornithomimid.

Ornithomimid: Wait, I need more time. I'm going to adapt like the green lady said.

Red: It matters not as the time has been set. The Earth does not change for life. Life must change for it.

Ornithomimid: Uh, I can still run. I can adapt. (runs away)

Red: You cannot outrun your death sentence.

(Red throws javelin and kills Ornithomimid)

Ornithomimid: Ahhh!

Red: Now what else is there to kill?

Bird: Uh, excuse me. Are you the blood spirit lady that is supposed to destroy everything?

Red: Yes I am, small dinosaur. How would you like me to take your life? It must be quick because I have lots of other things to demolish today.

Bird: Oh no blood goddess, please don´t kill me. Mother Nature sent me here to be of service to you. She gave me her seal and said that you would recognize it and spare my life.

Red: I was sent to destroy and kill everything in my wake. I'll paint it all red and blot it all out so that everything can start again on a blank canvas. Tell me pipsqueak, why does the green maiden want me to spare your miserable life?

Bird: She said that the seeds of life have to be preserved so that animal life can recover after the apocalypse.

Red: This planet has become nothing but an ugly garbage heap, overrun with decadence and corruption. There is no beauty to be seen in this mess. I must restore silence with violence and bring peace to this place, while the dinosaur blood flows. (Red prepares to kill bird)

Bird: Wait! I know what has to be done and I will help you in your mission.

Red: Hmmm. Perhaps your frightened soul may be of some use to me. Tell me where all the creatures of this desert are hiding and I will not kill.

Bird: Tweet! This land is barren. The large dinosaurs live on the plain in that direction. But there is one large flying reptile that lives on the cliffs by the sea.

(Red points spear at bird, bird trembles in fear, red grabs bird in hand)

Red: Where?

Bird: Uhhh, its nest is on the top of that spire of rocks over there.

Red: In return for not killing you, you must do something for me. Fly up to the winged beast's nest and destroy its egg so that its kind cannot recover after I leave.

Bird: I'm not sure I can do that.

Red: No, I need you to do this for me. If you do not do this, I will hunt you with my exploding grenades and poison you with their toxic vapour. But if you do do this for me, I will honour the green maiden's promise to you, and you will inherit the skies of the world

Red: (squeezes bird) Scream!

Bird: Tweeeettt! Let me go! Let me go!

Red: Bring that flying reptile to me.

Bird: Tweeeettt! It hurts, it hurts.

Red: Look, here it comes.

Pterosaur: Awk! Another one of these strange bipeds. Stand down creature. Release my underling for today it shall be my prey.

Red: Silence winged vermin. You shall have no prey today. You shall meet your end by the tip of my spear.

Pterosaur: You do not scare me with your words.

(Pterasaur pecks at Red and flaps wings. Red hits Pterosaur with shaft of spear. Pterasaur flies, swoops down on Red and takes her spear. Red pulls out her bow and hits Pterasaur with an arrow. Pterasaur drops the spear) Ouch!

Red: Go now, while this beast is away.
Bird: Well, I definitely don't want to be beside you.
Red: Do it!
Pterosaur: Die biped!
(Red ducks, rolls, pulls out sword and cuts Pterosaur's leg)
Pterosaur: Ahhh!
Red: Die parasite! (shoots arrows)
Pterosaur: Ahhh! (flies toward the coast)
Bird: Oh, here is the pterosaur egg. It is as big as I am. (pecks at egg) Drat, the shell is too hard. I've got to find some other way to break it.
(Bird tries to push the egg over the edge of nest)
Pterosaur: See if you can hit me with your flying needles now!
Mosasaur: Mmm... smells like blood falling from the sky. It's a Pterosaur. (SNAP!)
Pterosaur: Ahhh!
Red: And so the reptile that learned to fly now swims in the belly of the sea.
(Bird picks up rock, flies above egg and drops it, egg breaks)
Red: Little dinosaur, have you destroyed that egg yet?
Bird: The task is done, Pterosaurs shall terrorize the skies no more.
Red: Excellent. (reclaims spear) Now come on. To the forest. There is much more blood to spill.

Act 4, Scene 3

Setting: An afternoon in the forest. Red kills dinosaurs and destroys their eggs.

Enter: Red, bird, mammal, snake, Dromaeosaur, Titanosaur, Pachycephalosaur, Hypsilophodon

Snake: Yeck! The forest again. I don't like this place. It's full of ugly plants and dinosaurs.

Mammal: Quit your complaining, snake. We have important work to do.

Dromaeosaur: Well look who it is. Tell me little ones, is there light on the other side of the tunnel?

Mammal: Yes there is, but you will never see it.

Dromaeosaur: Oooh, I'm so disappointed that I won't be ordered around by little miss environment. Have fun at your little survivor party.

Snake: Sss!

Mammal: Calm yourself, snake. That dinosaur is just being rude.

Pachycephalosaur: Dromaeosaur, why don't you just get your raptor hide out here and go back to the plain. You are not welcome here.

Dromaeosaur: Relax bonehead. I'm not bothering you. I'm just trying to convince these small creatures that their life is not worth living so I can eat them.

Pachycephalosaur: Don't listen little animals, your life is great. I envy being small and being able to hide and survive on smaller amounts of food. You should be so happy.

Dromaeosaur: If you want to be smaller, Pachycephalosaur, I'd be happy to eat any parts of your body that you don't need.

Pachycephalosaur: Why you! I' going to crush your skull. (charges at Dromaeosaur)

Hypsilophodon: Stop Pachycephalosaur. This is senseless.

Pachycephalosaur: I don't care. I don't care. I just want to bash its head in.

Titanosaur: Okay Pachycephalosaur, but just injure that little thief. I want to stomp on it myself for what it did to my egg.

Red: The day of judgement has arrived for you creatures of the forest. The blood maiden is here to enforce your destiny. (lights trees on fire)

Bird: The large dinosaurs are my cousins. The two small creatures, mammal and snake, are also servants of the green maiden.

Red: I see I have a lot of cleaning up to do. I will start with those bulky tree-eating dinosaur ignoramuses. I'll cut them open and let their blood spill onto the earth.

Hypsilophodon: Hey! That doesn't sound nice. I'm sure you don't have to resort to that. I believe there is good in all creatures.

Red: (slices Hypsilophodon's neck) You are not needed here anymore.

Dromaeosaur: Alright, I'll be going to the cave now. (exit)

Titanosaur: Oh no you don't.

Pachycephalosaur: You killed Hypsilophodon! You monster!

Red: What about you, creature? Will you die by my spear or by my sword?

Pachycephalosaur: You shall die by my helmet. It's ready to crush your body beyond recognition. (Pachycephalosaur charges at red. Red dodges and turns, slicing Pachycephalosaur's neck as it runs by. Pachycephalosaur falls to the ground)

Red: Bonehead!

Mammal: That was pretty cruel natured of you.

Bird: Now the forest floor has seven tons of dinosaur meat to decompose.

Snake: I want to go back to the underground.

Mammal: Yeah, me too. I really don't want to end up on the wrong side of this sick fantasy of yours.

Red: No my servants. I will need you to do something for me. In return, you will have my respect. Your lives will be spared and you will inherit the top niches of these ecosystems. Seek out the eggs of the plant-eating dinosaurs. Show me where they are so that their bloodlines can't continue after I am gone. Their seed has to be extinguished from the Earth forever. Go now.

Mammal: I will follow my nose.

Snake: Sss!

Bird: Tweet!

Titanosaur: Where have you concealed yourself, Dromaeosaur. You cannot hide forever. You will have to surface sometime.
Red: En garde, long necked fiend. The time has come for you to face your fate.
Titanosaur: Away with you biped. I don't have time to discuss your environmental philosophies.
Red: Time is not up to you to decide. You remain at the mercy of the cosmic calendar. Prepare yourself for extinction.
(Red slashes at Titanosaur legs and body)
Titanosaur: Erghhh! Foolish animal. Not even Tyrannosaur can puncture my skin. (Titanosaur swings tail and Red dodges)
Red: Drat! My blade cannot penetrate the skin of this beast. (drops sword)
Titanosaur: I am a Titanosaur. One of the mightiest beasts to ever walk the Earth. It will take more than your steel fangs to hurt me.
Red: The Earth will find a way. It cannot sustain your decadence any longer. (fires arrows at Titanosaur's head)
Titanosaur: Ahhh, my eye!
Red: A fair action. You have been blind to your sins in life and so you will be blind in death.
Mammal: (squeak) It is done my lady. We have located the eggs of the forest dwellers arranged in a ring in the clearing over yonder.
(Titanosaur swings tail and hits Red, knocking her over)
Titanosaur: ARGHHH! The Titans rule. The ground shakes when I walk and I dominate the landscape wherever I go.
Red: (Sharpens spear with rock) No beast rules the Earth. It is the Earth that rules all beasts. (Throws grenade at Titanosaur, which explodes on impact)
Titanosaur: Unhhh!
(Titanosaur is disoriented and Red throws her spear at Titanosaur, piercing its head. Titanosaur falls to the ground)
Red: In time your body will be reincarnated into the bodies of plants and creatures that are yet to evolve. You, on the other hand, will be forgotten except in the annals of the fossil record. Now where are those eggs?
Bird: Over here.

Mammal: They are all in the clearing except for the bird's. It keeps it in the trees.

Red: Excellent. Then we shall crack the shells and scramble the yolks for an omlette of carnage.

Snake: Yesss! Let us feast.

Bird: No, not my egg. It's not in the trees, in fact I don't have any, I swear.

Red: Relax pipsqueak. I'm not after your egg. I will spare your offspring as long as you do one more task for me.

Bird: I am at your mercy.

Red: Let's scramble those eggs and let the yolks run. (slices and smashes eggs, leaving just a pile of broken eggshells) Our work is not done comrades. Follow me to the plain to finish this battle once and for all. (exit)

Bird: (gasps) Those were to be my little cousins. How can nature be so cruel? (exit)

Snake: Awww. I want to eat eggsss. (exit)

Mammal: Come on snake, stop your whining. (exit)

Dromaeosaur: Ahhh, the perks of being a scavenger. A meal tastes so much better when you don´t have to make it yourself.

Act 4, Scene 4

Setting: A late afternoon on the plain. Tyrannosaur fights Ceratopsian, Red kills dinosaurs, and mammal and snake flee to the underground.

Enter: Red, mammal, bird, snake, Ankylosaur, Ceratopsian, Tyrannosaur

Red: At long last I arrive at the plain, where the ground is stained with the blood of the dino warriors as they bring evolution to its most violent form. It is here that I will meet the apex predator and its competitors, and confront their oozing overconfidence face to face.

Mammal: You go ahead and do what you need to do, I'll just hide behind this rock.

Red: No furry one, today is the day that you must stop hiding and witness the dawning of a new age. For after the dinosaurs are gone, it is you who will inherit the Earth. You must realize your potential and learn to manage your home planet better than the dinosaurs. Keep your eyes and your noses open for dinosaur eggs. They must also be eradicated.

Snake: Eggsss!

Red: Look here! A giant armoured beast awaits its doom.

Ankylosaur: Ohh, I'm going to beat Tyrannosaur in the next match. This time I'll be quicker and stronger. My wounds are almost healed.

Red: En garde shelled fiend. It is I, Apogalacticon, slayer of dragons, here to carry out the will of the cosmos. Prepare to meet your destruction.

Ankylosaur: Who are you to say such words stranger. I know my end, it will be in the heat of combat as I battle my eternal enemy, not engaging with the likes of you.

Red: Silly vermin. You seem to have forgotten that extinction is your biggest enemy. You have overgrown your means with millions of years of decadent evolution.

Ankylosaur: No biped, it is you who have overgrown yourself with ideas of directing the course of evolution beyond your own organism and for that my bludger shall strike you down.

Red: Enough! (throws grenade)

Ankylosaur: Hah! My armour is four inches thick. Your magic is of no use here.

Red: We will see about that.

(Red advances with spear, stabs Ankylosaur, Ankylosaur swings, Red dodges, but loses spear, Ankylosaur stands on spear)
Ankylosaur: ARRR! How can you win if you lose your weapons so easily.
Red: (pulls out sword) Then a firm grip I shall keep as I slash you open.
(Ankylosaur stamps feet, turns and swings, red jumps and rolls right in front of Ankylosaur´s face, stabs sword into throat)
Red: So ends your saga, armoured reptile. Strong of back but weak of mouth. Rest forever, beast of burden.
Mammal: I wish these mass extinctions didn't have to be so violent. Can't you just put all the dinosaurs to sleep and let them down easy?
Red: No I cannot. I have to confront them face-to-face so that they hear the message loud and clear.
Snake: Look warrior princess, there is another dinosaur over there.
Tyrannosaur: (holding an egg) Oh sweet egg of mine, I will protect you and keep you safe from all harm.
Mammal: And here comes the Ceratopsian.
Ceratopsian: Tyrannosaur!
Tyrannosaur: Oh, here comes a plant eater. I'll put you back in my nest my precious baby.
(Tyrannosaur sets egg down into nest, looking very ridiculous of course)
Tyrannosaur: At last Ceratops, today you will finally be my dinner and my conquest of this dinosaur world will be complete.
Ceratopsian: You will not bragging when your head is on a spike.
Tyrannosaur: I'm going to give you a thrashing.
Ceratopsian: URMPHHH! I am going to kill you Tyrannosaur! (charges)
Tyrannosaur: Bring it on!
(Ceratops lunges into Tyrannosaur but his horns don't strike and Tyran bites Certops frill, they become locked in a stalemate like embrace)
Red: Hello monsters. Death has arrived for both of you today.
Snake: Blood mistresss, I senssse an egg around here.
Red: Go find it snakey. Now which one of you dinos wants to die first?
(Ceratops backs up and breaks away from Tyrannosaur's grasp, the snake and the mammal go searching for Tyrannosaur's egg, Tyrannosaur pursues Ceratops)
Tyrannosaur: No Ceratops, you will not escape. Nothing will save you this time.

Ceratopsian: Just try it big mouth. These horns have long waited to puncture your ego.
Red: Your bickering is pointless. Just let me end both of your pathetic existences and the will of the galaxy be done.
(Red stands in front of Ceratops, Ceratops knocks her spear and clashes with Red. Red eventually manages to poke Ceratops in the eye)
Ceratopsian: Owww! (Ceratops backs up, Tyrannosaur approaches Red)
Tyrannosaur: Who on earth are you? And how dare you try to steal the horned tasty from me. I am the master around here and only I will decide who gets to live and die. I will trample you where you stand.
(Red throws grenades but Tyrannosaur doesn't budge, Red holds up spear but Tyrannosaur bites it in half. Red pulls out sword and parries with Tyrannosaur's teeth. Tyrannosaur turns and swings its tail, knocking Red down)
Tyrannosaur: Ha! Dumb synapsid! Carnosaur big and strong. Knock you down to the ground. Me king of dino universe.
Snake: Sss. Here it is. A dinosaur egg big enough for all of us.
Mammal: I'll take it. (tries to pick up Tyrannosaur egg)
Tyrannosaur: Now feel the force of my foot. Surrender to my tyranny and let your body be my prize.
(Red rolls over, dodging Tyrranosaur's foot)
Red: You will be defeated, tyrant. If not by me then by my aftermath. That is your destiny. You cannot escape.
Snake: Get out of the way mammal. I'll get rid of this egg.
Mammal: No snakey. It's my prize. (starts to roll the egg)
Snake: No, it's mine! (eats Tyrannosaur egg whole)
Tyrannosaur: What? Hey, give me back my precious. Give it back now or I'm going to tear you open.
Snake: Mumble, mumble. (can't move because the egg is too big)
Ceratopsian: I'm going to run you down this time Tyrannosaur. Your blood will spill. (charges)
Tyrannosaur: Forceps! (steps on snakes tail)
Snake: Ahhh!
Tyrannosaur: Scalpel! (lifts up foot and extends claw)
Ceratopsian: Die Carnosaur! (horn punctures Tyrannosaur's leg)
Tyrannosaur: Ahhh!
Snake: Ahhh!

(Red picks up sword and runs up to Ceratops, stabs it in back)
Red: That will be the end of you hornface.
Ceratopsian: So be it. I shall die happy with my victory over the tyrant.
Tyrannosaur: Damn you all. My leg! (breaks away and regains balance) My egg! I'll cut you open.
Red: (cuts off Tyrannosaur's toe) En garde! Your daggers will fail at the blade of my righteous sword.
Tyrannosaur: Arrr!
(Tyrannosaur tries to bite Red, Red parries Tyrannosaur's teeth, Tyrannosaur lunges, Red dodges and slices Tyrannosaur's neck)
Red: There you go, dagger tooth. Your tyranny finally laid to rest. The era of reptiles has met its end today!
(Tyrannosaur makes death sound and falls to the ground)
Snake: (groans)
Red: And now serpent, what are we going to do with you. You can't go anywhere in that state. Hold still.
(Red steps on snake, shell breaks, snake pukes up egg shells and yolk)
Snake: Blehhh!
Red: Mammal, come hither and feast upon the remains of the tyrant lizard's young. Fulfill yourself and prepare for the journey back through the underground. Blessed are you, humble scavengers, for you shall inherit the Earth in the new era.
Snake: Uhh?
Red: You, serpent, will inherit the lower niches in the undergrowth and in the creepy crawly places of the world. Go now, both of you, to the refugium that awaits you. There, your lives will be spared from the wake of my destruction.
Mammal: Thank you, blood maiden.
Snake: Ssso be it.
Red: Now come on avian dinosaur, there is one more task to be done. Come with me now and leave your egg behind. You will have another chance for offspring in the new world and they will be free of this mess.
Bird: But it's the only thing in the world I really care about.
Red: It is a sacrifice that you must make. Despair not, for of all the dinosaurs, only you were fit for transcendence. Let go of what burdens you today and open your eyes to a new and better life.

Act 5, Scene 1
Setting: Sunset on the plain. Red creates a volcano, smoke and fumes fill the air, and bird flies over the mountains.
Enter: Red, bird

Red: Here, this will do. The shock waves of my impact converge here in the plain of Deccan. This spot shall become a blanket of lava unlike anything the Earth has seen in the last 100 million years. (Red stabs her sword into the ground and makes a large hole)
Bird: What must I do?
Red: Take a look around and bring me back any vegetation you find to help fuel the fire. You must bear witness to my masterpiece of destruction and bring a message to the green maiden that the task is done. (Bird goes flying in search of plants)
Red: Here, I place all of my unspent ammunition. It will spew forth a fountain of death. (Red drops grenades into hole, except one)
Bird: Here you go, I gathered all the scraps of vegetation I could carry.
Red: That will do, bird. Put it around the pit like a nest. Any creatures that are not in the refuge, and have not already perished by my sword, spear or bow will not survive the toxic wave of death that is coming their way. There, that is perfect. You have proven your worth for you are humble, you consume less and aren't burdened by the accumulation of excessive material bulk. Your ease of movement and ability to adapt to the changes has allowed you to transcend the fate chosen by all of your dinosaur cousins. Go now. Fly quickly to the refuge. You will be safe there.
Bird: Thank you for not killing me. I'll go and tell the others. (exit)
Red: Now let the Earth burn and the sky turn grey. Let death rain from the skies as the full influence of the cosmos is manifested. (Red drops grenade into the pit, which explodes. The Earth shakes and rocks are hurled into the sky. Smoke fills the air)
Red: Let the fire burn and the cauldron bubble. Let my blanket of death sweep over this planet. (exit)

Act 5, Scene 2

Setting: Evening in the open ocean. Mosasaur boasts. Then the sea turns dark and toxic and Mosasaur dies.

Enter: Mosasaur

Mosasaur: I am so great. The ocean belongs to me, the most powerful marine predator of all time. I have scared away that pesky shark, I ate the Elasmosaur and caught the flying reptile in my jaws. Oh and look, the dead body of Archelon is floating by. I'll just have to eat that too. (CRUNCH, CRUNCH, CRUNCH) Hmm...well, now that I have the ocean to myself, I might get kind of lonely here. No, never, the emptiness of this place will always stand as a trophy to my triumphant conquest of the ocean and my limitless power. Ha ha ha! But what will I eat now when I get hungry. There's nothing left. I ate it all. Agh! Don't be silly, that doesn't concern me. I'll just move onto land and eat everything there. I will rule the whole earth and nothing can stop me. I am invincible.

(suddenly some rocks fall through the water and the sea turns murky)

Mosasaur: (coughs) What? What is this? Ash and dust in the water (cough, cough, cough) How can this be, the very ocean is turning on me in my moment of glory. (cough, gasp) To the surface, the surface.

(Mosasaur swims to the surface, looks into the air and can't see anything)

Mosasaur: (coughs) No! (cough, cough)

(Mosasaur dies)

Act 5, Scene 3
Setting: Evening at the cave. Mammal and snake go through the underground passage. Spider stings Dromaeosaur.
Enter: Dromaeosaur, mammal, snake, spider

Mammal: Where is the passage to the lake?
Snake: Sss, this way.
Mammal: Okay.
Snake: Ssstop, I sense sssomething.
Dromaeosaur: Ah, the legless lizard, we meet again.
Mammal: Your time is up, carnivore. Get out of our way.
Dromaeosaur: Oh it's the furry freak. Tell me where that refuge is.
Mammal: You're a dinosaur. You can't go there. You have abandoned mother nature and in return mother nature has abandoned you.
Dromaeosaur: You know I would eat you if I could find you in this dark place.
Snake: Sss! You'll never find me. (sneaks past Dromaeosaur)
Dromaeosaur: I hear you snake, I hear you.
Mammal: Raptor, leave the snake alone, you are not meant to be here.
Dromaeosaur: You little creatures are wasting your time with fantasies. Face the reality, you are my prey.
Mammal: I am no longer your subordinate and will believe your lies no more.
Dromaeosaur: Just keep talking. I'll find you soon enough.
(mammal throws a rock in the cave and then sneaks on the other side of the passage)
Dromaeosaur: Gotcha! (bites rock) Yeck!
Snake: Sss!
Spider: Whoah, it's you again I suppose you want me to let you through.
Snake: Yesss, spider we have to get back to the refugium, remember the deal?
Spider: I want more bugses.
Snake: I don't have any to give you, I'm going to the surface, you should think about doing the same.
Spider: Oh, no you don't. (bites snake)

Snake: It's no use spider, I am venomous like you. You can't stop me and neither can your web. I'm just going to slither right through it. (exit)

Spider: Arghhh!

Mammal: Spider, let me pass. We have to go to the refuge right now. There's no time to waste.

Spider: Don't order me around. Give me some insects or I will poison you.

Dromaeosaur: I can hear you mammal.

Spider: Say your last prayers furry one.

Mammal: I need to get past your web.

Dromaeosaur: So mammal, you tried to trick me. You'll pay for that. (snatches at mammal with claws)

Mammal: Eek!

(mammal moves out of the way and Dromaeosaur cuts through the spider web with its claws)

Spider: Ach! You idiot dinosaur. That's my web. Do you know how long I worked on that?

Dromaeosaur: What is this stuff?

Spider: Taste my venom! (stings Dromaeosaur)

Dromaeosaur: Ouch!

Mammal: So long raptor.

Dromaeosaur: I feel numb and I can't... (Dromeosaur becomes paralyzed and faints)

Mammal: Towards the light we go.

(mammal goes through web)

Spider: I'm going to sting you too. (follows mammal)

Mammal: That's it, come on, this way. (exit)

Spider: You coward, I will hunt you down like the vertebrate scum you are. (exit)

Act 5, Scene 4

Setting: Evening at the freshwater lake. Green and Blue speak to survivors.

Enter: Green, Blue, mammal, bird, snake, frog, lizard, Crocodile, lobster, ammonite, fish, Shark, spider

Mammal: At last, the sounds of running water and insects buzzing and the sweet aromas of flowers surround me. I know I'll be safe here.

Green: Excellent! Our furry friend, the mammal, has arrived. Tell us of the battle with the dinosaurs?

Mammal: The dinosaurs are all dead now, except the bird, of course.

Lizard: Good riddance.

Crocodile: They were my favourite food. What am I to eat now?

Bird: The blood maiden sends her message Mother Earth. The task is done.

Green: That is great news. Rest and be safe here my servants. May the solitude bring peace to you.

Spider: Ahh! The light! What is this place? Where is the mammal, that furry runt?

Green: Eight legs! At last you have arrived. Welcome to the pond. Here you will find relief from the blanket of extinction that is covering the rest of the Earth. We are happy to have you with us. You may help yourself to whatever insects you can catch.

Blue: (distributing shellfish around the lake) To all my swimming friends, blessed are you for making the journey. Stay here for a little while while the ocean recovers. The next era of marine life will eventually belong to you.

Fish: Hooray! Thank you Siren for saving us from the apocalypse.

Lobster: So you made me travel all the way up the stream with my heavy shell and now you're saying I'm going to have to go all the way back.

Ammonite: You took my shell from me. I want it back. Look at me, I'm just a naked tentacled beast. What am I going to do now?

Shark: You made me suck on this salty rock and man I just feel so ridiculous.

Blue: Forget your troubles for now my friends. In time you will realize the rewards for your sacrifices. And you, swamp creatures, who are already quite comfortable in the freshwater, your life shall continue much as it was before.

Frog: Ribbit. Ooh, I can't wait to get back into that dirty, smelly swamp.

Green: Now my creatures, big and small, from the tiniest insect to the crocodile, you all play a vital role in the web of life. Go now and live your lives again. Be fruitful and multiply. The Earth is yours. Let your offspring roam all parts of the world. But remember our words and remember the dinosaurs for in 150 million years, Apogalacticon, the galactic summer solstice, will strike again and your descendents shall be tested as you were.

Blue: Farewell my brave followers. You have proven your worth to the Earth. May you find happiness in the days to come. (exit)

Green: Farewell little ones (exit).

Mammal: Farewell

Bird: Bye Mother Nature. I will always remember you.

Fish: Hooray, we're safe.

Shark: Not since that witch is gone you're not.

Fish: Yipes!

Frog: Well, I guess that's nature for you.

THE END

Manufactured by Amazon.ca
Bolton, ON

39331067R00035